When Every day Is a "New Day"

Author Marlean Acker

WHEN EVERY DAY IS A "NEW DAY"

A LIFE LIVED WITH DEMENTIA/ALZHEIMER'S

Marlean Acker

Published by Living Water Book Publishing LLC

Little Rock, Arkansas 72201

Livingwaterbooks.org

Print Book Edition 2022

Library of congress cataloging in publication data under

ISBN 979-8-9868286-4-0

Graphics, Art, and Designs Living Water Books

In Loving Memory Of Momma

Barbara Jean Williams

Mays-Sumerlin

Acknowledgments

This book is dedicated to the thousands of people who have been caregivers for loved ones battling Dementia/Alzheimer's.

My mother, the love of my life
Barbara Jean Mays-Sumerlin (Williams), her struggle was real.

Special note to my loving dedicated husband
King David Acker, who was Mom's primary caregiver.

My family, especially my great sister Christell O'Neal, thanks so much.
Humble thanks to Rita Deloney, "Deloney's Adult Daycare."
Rita understood Mom and did a wonderful job working
with Mom's disease and personality.

I cannot leave out our Promise Land Church Family,
under the leadership of Dr. A. Marquis Scruggs Sr.

Our lives have been challenging, love-filled, eventful, refreshing, fantastic, and loving having taken this journey. I asked God to bring my mother back to me while she was still able to recognize me. God did it! I love Him for it!!!

Table of Contents

Introduction

The journey you will take reading this book is so well worth the trip. It chronicles the life of an amazing woman, mom, grandma, great-grandma and friend. She reared two successful children on her own after a divorce.

Barbara had a full life before the gradual, devastating effects of Dementia/Alzheimer's.

She laughed, loved, cared, traveled, married again, worshiped, and served.

This book will show you her life as the disease progressed. I want you to be encouraged as a caregiver that there are going to be some rough times, but the good will always outweigh the challenging times.

Journal your way through the experiences of this book. Enjoy.

The Canvas Of Her Life Before

Mom was a loving wife to a military man, Leon Sr., and a dynamic mother of two children: Marlean and Leon Jr. She was a licensed beautician, and in the early years, she simply did hair at our home. She later took a manicurist job with Leigh Jones, which lasted over 30 years. She was full of so much fire playing in at least six bowling leagues a week. She actually bowled a perfect 300 game in her lifetime. She traveled all over the country to bowl in tournaments which brought her great joy. She worked hard to put us in private school, also working at night in her cousin Joe's bar to make ends meet.

By the way, my mom never smoked or drank.

When I was four (around 1960), my mom became a single parent. She also did hair on the side for friends and family and went to clients' homes to do their nails and feet. She worked very hard to make sure we had every advantage as it pertained to education.

She worked into her 70s until her memory would not allow her to continue. She loved to play cards, was a member of a Bridge club, and played bid whist with a fierce competitive spirit and passion.

She did remarry the late Norman Sumerlin in 1968. He was a police officer. They were together nine years before infidelity on his part caused the marriage to crumble.

Mom took it in stride and continued her quest to make sure she provided for all of us.

I went to college on a full academic scholarship, and my brother went to the Air Force. He returned, and I graduated from Roosevelt University, an extra proud day for my mom. I applied to go to Officer's Training School with the Air Force and was accepted.

This left mom alone, enjoying life.

My brother was living downstairs with his girlfriend and four children. Mom was upstairs with the miniature schnauzer I sent her from New Mexico.

I DESCRIBE HER AS VIBRANT, HARDWORKING, FIERCE, AND DETERMINED TO SEE HER CHILDREN SUCCEED.

One dark day the building caught fire and destroyed the entire house; both families lost everything but were physically ok.

My brother was missing, then we got a call from the police, and they informed us that my brother was found hanging from a bridge.

My mom had just gone over that bridge when he was found. I didn't know what was happening. If she had, I could have lost them both. Mom's blood pressure was already sky-high (no medication due to the fire.) I took her to the store to get some personal items. She was in a daze. A church took us in, and funeral arrangements were made.

My father contacted us and asked if we needed help.

Yes, we stated, and he promised to do so.

At the funeral, he did not come through at all; he just lied.

My youngest brother, whom I had never met, spoke; my grandma, aunts, and uncles came. I had not seen them in over 30 years.

Life goes on.

I left New Mexico and returned to Arkansas.

Mom was able to get settled in an apartment, and my brother's girlfriend Tanya and the children lived with her. Tanya was low functioning and did not work. The children were toddlers and received Social Security.

Mom thrived, helping to take care of those grandchildren and Tanya too. I don't really know how Mom got off track with her medications and blood pressure, but it did take a toll that Doctors did not seem to recognize early.

I truly believe this had a lot to do with her onset of memory loss.

She started to decline about five years before. I noticed it on her visits to see me in Arkansas from Chicago.

And so, it begins ...

Chapter One
Where We Began This Journey

First, let me tell you that the warning signs of Mom dealing with forgetting (Dementia/Alzheimer's) were manifested years before the beginning of this book.

She was writing herself notes to remember things when she returned home after visiting us in Arkansas. Once we put her on the plane and let her know that her granddaughter Marlean would pick her up from the airport. She totally forgot. We thought she had been missing for hours. We found out that she rode the commuter train home from the airport.

In August 2009, we received a call from a stranger long distance from Chicago.

She says, "You are going to have to check on your mother; she is in the hospital. Tanya is also."

Tanya was my brother's children's mother. At the time, her four children ranged in age from 16-21, and she had one grandchild,

11 months old. They had lived with my mom since 1997, after my brother, Leon Jr., passed away.

"What?" I asked frantically. "Where is she?"

I had so many questions.

I called my daughter and gave her the details, telling her that Mom was at Loyola Hospital. Then I called the hospital to notify them that no one was to pick up or remove my mother from care except my daughter Marlean and her Husband Milton (Bookie).

Marlean made contact with the hospital, and we got more details.

They found Mom on the streets with no shoes or purse. She had an apparent head injury either from being mugged or a fall.

We still don't know the whole story.

Medically she was doing fine; however, it was obvious she had gaps in her memory and was starting to deteriorate.

THE NEXT DAY

The hospital transferred Mom to a nursing home.

Mom escaped and was found walking up the street, attempting to return home. A nurse convinced her to return. Well, she escaped again and went to the hospital and made a terrible scene, cursing and yelling at the staff, demanding to see her doctor.

She required a police escort back to the nursing home.

Marlean and Bookie picked her up from the Nursing Home and took her home with them until I could get there from Little Rock, Arkansas.

Bookie was the one who stayed up all night to make sure Mom was secure and did not get out of the house. He gave her medicine and kept her on schedule. Bookie was so kind and patient with Mom. (She responds better to males. We'll talk more about that later.)

She still managed to get out of the back door of the apartment.

Bookie followed close behind and convinced her that she was too far from her home to walk by reminding her there was no bus that ran out there and pleading with her to come back to the apartment.

Mom had to be watched closely to keep her from getting out of the apartment.

She was so determined to go home that it was a constant and continuous struggle to convince her to stay. My son-in-love was a life-saver. He was able to keep an eye on Mom, and when things went out of control, he was able to calm the storm

Mom always responded positively to young men, and we used that to our advantage in this case.

Chapter Two

Another Day

Upon our arrival, Mom was very confused about where she was and kept demanding to be taken home. She was very frail, about 113 pounds, soaking wet.

She probably forgot to eat properly.

During this same time, Tanya passed away.

We debated whether to take Mom to the funeral considering her fragile and confused condition. However, we decided to take her and learned she really did not realize who had died.

We tried to keep her from knowing who had died, but she got a copy of the obituary anyway.

It was so sad that Mom was not mentioned.

"I'm a grandma," she said. "Where is my name?"

They had not included her in the obituary.

I had to get up and speak for her, setting the record straight:

My mom was dedicated to Tanya and those children (four) for over 14 years, working a job well into her 70s until her memory loss precluded her from continuing to work.

Tanya's family did not assist her in any way during the whole time. Mom was really hurt, and I spoke up for her during the service.

***

After the service, we pretended to take her home and just kept driving to Little Rock, Arkansas.

Mom just rocked in the back seat holding a pillow and mumbling incoherently.

We got out to go to the restroom one time.

"I've lost my car keys and my purse," Mom insisted.

"No, Mom," I said. "You haven't. I have them here with me. Please get back in the car."

Convincing her of that was no easy task, I must say. But finally, she got back in the car, and we continued our journey to Arkansas. When we arrived, my God-sister, Christell, was waiting dressed in her scrubs to help us with Mom.

We thought that would comfort her.

It had the opposite effect!

Mom ran from Chris, and we had to send her home to change clothes.

Finally, we got mom settled in her room and met to get a strategy to secure our home.

Mom was tired. She slept a little while and then had something to eat. Then she began asking to call home so she could check on the kids and Tanya. She wanted to let them know where she was.

Tanya's children were 21, 20, 18, and 17 at the time.

We assured Mom that they knew where she was and that everything was ok.

"And your phone isn't on."

“Yes, it is,” Mom insisted. “I paid the bill.”

“No, Mom,” we said.

We dialed the number for her and let her listen.

She stormed off to her room and came back 10 minutes later with the same discussion. We decided just to ignore her and change the subject.

“I need to use the phone,” Mom said.

“We don’t have a phone,” we said, hiding our cell phones.

She thought the TV remote was a phone.

Oh boy, that conversation went around and around for a while. She was relentless and was not going to give up at all. When we told her she was safe and that the children knew she was with us in Little Rock. She refused to believe that she was in Arkansas with us and that she had just lived around the corner.

This was a never-ending conversation, and we could never convince her otherwise.

ANOTHER NIGHT-SUN DOWNING

Sleepless nights soon became the norm for us.

I was so tired because Mom was very active at night, moving around the house and trying to figure out how to get out.

One night, around 2 am, we saw mom running up the hall with two blouses in her hand.

"I got to get out of here. My money is gone, and I can't pay for this Hotel."

Mom spent a lot of time in hotels and going to bowling tournaments in her past. She was mistaking my home for a hotel.

"Mom go back to bed," we told her. "You are not at a hotel. You are at our home."

We would get her settled back down, but every night it was more of the same. For two weeks straight, we dealt with Sun Downing episodes and the arguments about where she was and why. Sun Downing is manic behaviors that began at sun down.

HELP!

***

One day before we installed the deadbolts, Mom moved a solid wood table away from the door like it was plastic. She could not open the latch, which was the only thing that kept her from getting out of the house.

Chapter Three
A Different Day-Doctor's Visit

We got Mom a physician. Dr. Akins at the UAMS Longevity Clinic. At our first appointment, Mom was looking over my shoulder at her paperwork:

"That's not my address," she barked at me.

"No, Mom," I responded, "we use my address while you are with me."

She just shrugged and said, "Okay."

While I'm giving the nurse an update, Mom chimes in, "Why are you talking for me? I can talk?"

"Okay, Mom," I say calmly, avoiding conflict.

Later, I pulled the Dr. aside to tell her what I had to say about Mom's health.

On our next visits, Mom was a lot more cooperative and allowed me to converse with the nurse and the Dr. about issues she was having.

Her blood pressure was still high; we were trying to figure out why she ended up changing her medicine.

Mom is really starting to adjust well. (sigh)

***

One day, I came home from work, and Mom whispered to me that my husband had two young ladies in the house.

I yelled to the front room, “Babe, Mom says you had two young ladies in the house.”

“Shhhh,” Mom said, “Don’t be telling him what I said.” Then she just stormed out of the kitchen to her room.

Me and King laughed so hard our sides hurt.

***

Mom was quick too. She would sit at the breakfast table looking to see if we remembered to lock the deadbolt. I ran out one morning and forgot.

King fed Mom and went to the restroom.

On his return, she was gone, and the door open.

He ran up the street with a cast on his foot to catch her, pick her up, and take her back home. He thought for sure I was going to have to bail him out of jail or find him up against the wall because the neighbors called the police and told them he snatched up a little old lady and took her into his house.

SOME OTHER DAY

A friend suggested we have Mom admitted to the hospital to get her medicines balanced and get her to sleep more hours a night.

We did. It also gave us a chance to rest - although we visited her in the hospital every day.

They were able to get her up to seven hours of uninterrupted sleep before she was discharged. The only incident was her trying to shave with a plastic knife and messed up her chin.

They also prescribed her a patch to help with dementia.

She was pretty calm in the hospital.

I hated to leave her. She looked so sad and fearful that we were not coming back for her.

This was a needed break for us to get her medication regulated for us to administer to her at home. It was light night and day with her now.

We took her home, and peace be still!

We were all sleeping now; thank you, Lord!

We could tell if she managed to remove the patch or if it came off because her behavior would revert, and the first thing I'd do would be to look for the patch on her back to make sure.

Her blood pressure medication was changed to help keep it under control. Mom began eating regularly and even gained weight.

I had to buy her new pants every week!

I didn't mind. For me, it was a good sign – a sign that she was getting healthier.

We installed a key deadbolt on the door and had peace of mind when we slept at night.

Works really well if you remember to lock it.

We'll talk about that later too.

Chapter Four
Let's Start Daycare

We finally got approval from DHS for Mom to go to an Adult Daycare part of the day. This would give King a well-needed break during the afternoon. We chose afternoons because my mom has never been a morning person. She was in the beauty industry and never scheduled early appointments for her clients.

I would come home at lunch and take her to the facility and pick her up by 5:00 pm each day. I introduced her to the owner, Rita, and got her settled in. I assured her I would be back to pick her up.

Mom had a fear of being left in a Nursing Home.

She would say, “Don't forget where you left me.”

The first couple of days, she suspected that I wasn't coming back for her. So, she really looked surprised when I arrived at 5:00 each day.

Ms. Rita had to make sure the door was locked, and the key was around her neck since she had more than one escape artist (including my mom).

Ms. Rita gave me a report every day about things going on with Mom and whether she was eating and socializing. She decided it was not a good idea to let Mom play cards. Mom was an avid bid whist player, and she was way too serious and intense for the little old ladies in the daycare.

So, no card playing for her at the daycare.

WEEKDAY CHURCH SHENANIGANS

I would pick Mom up from daycare and go back to the church to finish work. At our 65th street location, Mom found her way out of the building.

She refused to come back with me. She said she was mad at me. The police arrived and would not let me take her. I explained the issue to them (Dementia / Alzheimers) and instructed them to ask her what city she was in to prove my case.

She answered, “Maywood, Illinois.”

To this, the policeman told me that I needed to have some kind of paperwork on her. A co-worker she trusted helped us get Mom back to the building. That was our location on 65th St. In our new location on University, she would ask to go to the restroom and insist that she knew the way to the restroom. However, she would be walking the other way and did not want to be corrected at all.

She would eventually get there.

***

We would find her trying to follow Pastor Chris King and his wife Janice out. They would peep in our office door to let us know we were the last in the building and that they were leaving.

Mom would jump up and say, “I’ll go on home now too.”

Oh no, here we go.

So, the Kings started just leaving without telling us so mom would not be triggered to leave with them. We bought some small chains to secure the doors, so we did not have to worry about Mom leaving the building on her so-called restroom breaks.

She would yank on the doors, realizing she could not get out and just return to the office.

***

I had to make a big note for my desk.

Mom would sit and do her puzzles. After a while, she would be looking for her purse to leave.

The note read:

> Mom, your purse is not lost. It is at home.

Wednesday night bible study was also a challenge at this location. Once Mom walked out and sat on the bench near the door. She said she was cold and trying to warm up.

The next thing we knew, she was walking away.

A church member followed her.

She got to a ravine, turned around and saw the church member behind her. They walked her back to the building.

Mom turned around and said, “Don’t be telling my business, either.”

Chapter Five
Mornings Are Always Fun

Mom was not a morning person.

We would wake her up for breakfast, which was steaming hot and ready for her. She would take her time getting up and dressed then her food and coffee would be cold.

She would then fuss and yell, “My food is cold.”

“If you came when I called, it would still be hot,” King would tell her.

Mom also liked a lot of sugar in her coffee and food. We substituted it for Splenda. She hated it but eventually got used to it. We changed to packets so she would not use too much, making her coffee and oatmeal bitter.

It’s the little victories we celebrate!

Sometimes, you would think. Mom was just plain mean.

We had a cure for that, and you would never guess what it was.

Ok. Ok. I am going to tell it. It is chocolate!

She could just be cursing you out, about to slap you, but if you broke out a candy bar, she would squeal like a giddy child and jump up, trying to get it out of my husband's hand as he held it up high.

It was a sight to see.

***

I accidentally gave her a Valentine's Day candy when I picked her up from daycare. There were about six pieces of chocolate inside of a cute little heart-shaped box. Before we got home (five minutes), she had eaten every piece of that candy!

Oh, did I mention my mom is actually allergic to large amounts of chocolate?

Oh my, we panicked, but she did okay - no emergencies.

From that point on, we knew we had to ration the candy and even hide it from her up high somewhere else in the house.

THE ACADEMY AWARD "OSCAR" GOES TO "BARBARA JEAN"

Once Mom demanded that we call an ambulance for her.

"My chest is hurting," she cried. "I need to go to the hospital! This life-threatening chest pain," she insisted.

I gave her two Tums and an Alka Seltzer and did not hear any more from her that evening.

We also put a call into her doctor and got a prescription for acid reflux. We also changed her coffee to decaffeinated and limited how many cups per day she was served.

***

We used to take Mom to church but had to monitor her trips to the restroom. She chose to cut up in front of others and fall out like we had hit her.

The award-winning academy performance by a lead actress goes to - Barbara Jean.

No, that was not the only award-winning performance of the year.

She always tried to leave the church by saying she had to use the restroom. We would catch her, gently grab her arm, and ask her to return to the office. She would see a group of people and just fall out on the floor, making it look like we pushed her down.

No ma'am.

"Please get up, Mom," I'd plead, "so we can go to the office.

We asked the people to kindly not intervene when we were handling her. She would eventually get up and go with us to the office. I am so glad no one tried to report us for senior abuse!

***

Mom would always find a way to get my husband's puzzle books. She had her own with her name on them. We wondered how she was finishing her books so fast. We looked in one of them, and she never finished a puzzle. She'd forget where she left off and instead move on to the next puzzle! When we told her his books were not her books, her performance could definitely have won an award!

Chapter Six

From The Lips OF Friends

One friend told us about a father who had Dementia/Alzhemizer and was really determined to get out of the house. He took a tablespoon to the wall of his closet until, at 3 am the next morning, he broke through the wall and got out of the house!

His family had no idea he was gone until morning.

A client of my sister Chris's was totally bedridden and did not walk. His wife was in another part of the house, and when she returned to his room, he was gone. She was frantically looking and wondering, *how could this even be happening?*

She looked out the front door, and there he was, sitting on the porch in a chair with his legs crossed.

To this day, no one knows *how* he got out to the porch!

***

My co-worker in the tax office said her mother was so determined to get out of the house that she would break out of the front bay window. She never cut herself and did this at least three times before they moved.

***

Another one of my co-worker's 94-year-old mother lived with her sister. My co-worker agreed to watch her mom while her sister ran errands. She had to run out to the car to get something. She did not have keys, so she told her mom not to lock her out of the house, which she promptly did. She also would not open the door or respond to phone calls.

She decided to check the backyard gate, and thankfully, it was open. There was her mom outside, watering the plants!

She asked, "Why did you lock me out, Mom? I've been calling you to let me in."

"The phone hasn't rung," her mom insisted. "I have the phone right here in my pocket."

She then pulled the TV remote out of her pocket to show her.

***

Unfortunately, sometimes, the stories aren't so funny.

Another friend told of how on a cold day, she was visiting her mom in the nursing home. However, on this day, her mom was determined to go home (the one she remembered). She managed to escape from the nursing home. A silver alert was sent when it was discovered she was gone from the facility. When she was found, it was too late; she had frozen and perished.

Yes, dementia has negative, tragic circumstances too.

***

Our close friend's dad was well known and loved around College Station. He checked up on him daily, and for some reason, on this day, he could not reach him. He called a couple of his friends to no avail.

It was time to start looking for him.

Everyone helped. They searched for days as panic started to set in for Tommy. Unfortunately, he was finally found under a bridge by the ravine. He had died of hypothermia.

A sad and tragic end.

***

Chapter Seven
A Flirty Day (Mom Loved The Fellas Forever)

Our Pastor, Dr. A. Marquis Scruggs, was a nice-looking, 40ish-year-old man at the time. Every Sunday, when we left the church, he and his wife Danita got a hug instead of shaking hands on the way out. Mom would then get back in line to get another hug from Pastor. Pastor told Mom she needed to stop before his wife started to get concerned.

***

Mom also had a hard crush on the church custodian, Mr. Green. She would light up every time she saw him anywhere. The week she passed away, he came to visit, and she perked up enough to entertain his teasing.

"You ready to go out," Mr. Green teased.

"No," she declined with a smile. "Not this time."

Mom loved talking to young, good-looking guys.

She used to spend six days a week at the bowling alley – she knew how to 'kid around.' She would have those guys blushing. We never knew what she was saying to them, but they took all in fun-loving on Mom.

***

One time, our Goddaughter, Tasha, was doing my hair, and her husband was waiting on her in our living room. Mom was sitting with him, holding a conversation. He was wearing a sweater and got hot, so he pulled it off.

Mom said, "Let me get a good seat. He is going to take off some more."

Byron just blushed at the thought.

***

At Mom's memorial service, Brother Lester Mckenzie spoke. He loved my mom and would make sure to get his hugs from his *grandma figure* every week.

"I thought I was the only one getting all those special hugs," he said before singing, "Grandma Barbara was hugging a lot of other people too."

Chapter Eight
Days Of Vacation With Mom

A couple of years into this journey, we decided to go on vacation to New Mexico to visit the children and grandchildren. Mom traveled okay. She did get sick a little, too much coffee. She really enjoyed the trip, and we had no issues or concerns with her wanting to leave; I guess she recognized being on vacation.

Her grandchildren went overboard giving her coffee, and she got sick again. Although telling her "No" was the correct answer, they did not like doing it.

It was a good trip.

We had a chance to take Mom to some museums. The Mosaic Temple Museum had a display about Dunbar H.S. mom remembered she went to Dunbar Vocational school for Cosmetology. Her long-term memory was very sharp. However, she couldn't tell you what or if she had breakfast. She could even recall people and places, and events around her High School years which I found to be amazing.

She really enjoyed the museum and all of the history. She was able to share a lot of her memories with us, such as when she got her license, her first job, and more things I really never knew about her education.

Wow, it is amazing what things can spark memories in people with her condition. She could still perm hair. Mom would do my hair for me, and it was so straight when she finished that I couldn't even get it to curl!

She was the consummate professional.

THIS DAY OH LORD

WE JUST NEED A BREAK DAY

We took mom with us wherever we went. So, sometimes we just needed a break. Our sister, Christel, would sit with mom at her house.

Their visits went well for a couple of hours - talking and eating, sitting in her favorite lounge chair. She would always fixate on a portrait of King and me every time she visited Chris's house. She repeatedly asked, "Where did you get that picture from? Why don't I have this picture?"

As time passed, Mom would get restless and ask when I would be returning and if it would be long. Chris locked her front door with the key and put the key in her pocket so Mom could not get out of the house.

When we arrived to pick her up, we found that all was well. Mom was ready to go 'home.'

Chris's daughter would take Mom for an evening sometimes as well. She had a harder time controlling Mom's behavior (For example, Mom trying to leave etc.). She had to work a little harder to keep her entertained and my grandson, Darius, was so good with Mom. He helped a lot. We just had to be creative when needing a break. Thank God for friends and family support. Don't fear asking for help.

Chapter Nine
Bath Day In The City-Then We Eat

Bath time was always a challenge. My mom did not want to get cold. We had an old fashion heater in the bathroom, and she still did not want to bathe. I figured out a successful strategy; I would run her water, turn on the heater, put down her towels, and simply announce to her that her bath was ready.

Then I continued to do whatever I was doing in another part of the house. I would always ask her to let me know when she was ready so that I could help her into the tub. A few minutes later, we would hear a splash of water, and Mom was already in the tub.

Success. Again, her way.

Although she fussed about having her back scrubbed (I had to get the adhesive off from her patch), I think she secretly enjoyed the scrub.

I guess the moral of this chapter is Mom will do 'anything' as long as it is her idea.

***

Those times 'in-between' meals were always exciting around our house. What am I talking about?

Glad you asked.

Mom would shout, "You ain't going to give me anything to eat?"

"Mom, you just ate," I'd say. "Don't make me take a time-stamped picture of you eating."

To that, she would say, "Shut up and don't be getting smart. I am *still* the mom."

THE DAY AT THE RESTAURANT

Restaurants, oh boy.

We went to Cheddars. The first thing we did was take Mom's picture with the handsome young host at the door. Yes, she really enjoyed that part. However, we had a wait for our table. So, Mom did not have to stand, I seated her with a nice young lady and her son.

When I went back to get her to go to our table, she got belligerent,

“Mom, we are going to our table now,” I quietly explained.

Mom finally understood, and we went on to our table without further incident.

Of course, by now her coffee was cold, which is always a problem except at I-HOP. There has her own pot of coffee.

We rarely took Mom to a buffet; she ate like a bird, and it was a waste of money. Plus, an hour later she would then accuse you of not feeding her.

Again, I threatened to take time-stamped pictures of her eating.

And again, she told me to stop getting smart with her.

“I am still Mom, you know?”

Chapter Ten
The Day We Took Vacation

We had planned a 7-day cruise before Mom came to us. So, I arranged respite care for her at a memory Nursing Facility. We got her settled in her room and brought her TV for her. We placed a note I typed and a picture of us both in her room. In the note, I let her know that we were on vacation. what day we left, and what day we would return to pick her up.

She looked sad, thinking we were going to leave her there forever. My mom was a true actress and drama queen. She was so genuine in how she felt but so dramatic with the delivery of the information. My sister, Chris, went to see her every day. She reported that Mom would barely eat, and she was so sad. She would NOT socialize with anyone in the facility. I was told she spent most of the time sleeping or just lying in her bed. However, upon our return to retrieve her from the facility, she perked up, and all was well with the world once we got home and she was settled back into her routine.

TIME MOVES ON

As Mom neared the end of life, Alzheimer's disease robbed her of her creative mind and memories, but her trust in God remained. She lived in our home for a season, where I was given a 'front row seat' to observe God's provision for her needs in unexpected ways, ways that helped me see she had been right all along.

Instead of worrying, she entrusted herself to One who promised to take care of her, and He showed Himself faithful.

Loving Lord, please help me to trust You to take care of me today, tomorrow, and forever!

Don't worry about tomorrow—God is already there.

James Banks

A Day For You

My message for caregivers:

Make time to take care of yourself. Indulge in a day spa, alone time, reading, or a good movie. Take a walk or run. Enjoy a nice night out for dinner. Make use of your support system (i.e., family, friends, co-workers, professional caregivers etc.) to avoid burnout and minimize your stress levels. There will always be another day, so you need to be mindful to think of yourself as well.

Self-care can be defined as a meaningful gift of self-love and appreciation. It is any activity we do to step back and rejuvenate ourselves mentally, physically, and emotionally.

SELF-CARE WORKBOOK

My Morning Gratitude Notes

DATE

3 THINGS I AM GRATEFUL FOR

I AM LOOKING FORWARD TO

I AM AT MY HAPPIEST WHEN

HOW ARE YOU FEELING TODAY?

WHAT WOULD MAKE TODAY GREAT?

POSSITIVE QUOTE

ACTS OF KINDNESS I WILL DO TODAY

SELF CARE ACTIVITIES

My Evening Gratitude Notes

DATE

BEST MOMENTS OF THE DAY:

PEOPLE I'M GRATEFUL FOR:

WHAT WOULD HAVE MADE TODAY BETTER?

POSSITIVE AFFIRMATION OF THE DAY

THINGS I AM GRATEFUL FOR:

3 ACCOMPLISHMENTS FROM TODAY

THOUGHTS AND FEELINGS:

Date: ______________

TODAY'S MOOD

SELF-CARE LIST

- ....
- ....
- ....
- ....
- ....
- ....
- ....
- ....

THINGS THAT MADE ME HAPPY TODAY

1. ....
2. ....
3. ....
4. ....
5. ....

AFFIRMATION

....
....

INSPIRATION

....
....

My Morning Gratitude Notes

DATE

3 THINGS I AM GRATEFUL FOR

WHAT WOULD MAKE TODAY GREAT?

I AM LOOKING FORWARD TO

POSSITIVE QUOTE

I AM AT MY HAPPIEST WHEN

ACTS OF KINDNESS I WILL DO TODAY

HOW ARE YOU FEELING TODAY?

SELF CARE ACTIVITIES

My Evening Gratitude Notes

DATE

BEST MOMENTS OF THE DAY:

PEOPLE I'M GRATEFUL FOR:

WHAT WOULD HAVE MADE TODAY BETTER?

POSSITIVE AFFIRMATION OF THE DAY

THINGS I AM GRATEFUL FOR:

3 ACCOMPLISHMENTS FROM TODAY

THOUGHTS AND FEELINGS:

Date: ______________

TODAY'S MOOD

☹ ☹ 😐 ☺ 😃

THINGS THAT MADE ME HAPPY TODAY

1. ....
2. ....
3. ....
4. ....
5. ....

SELF-CARE LIST

- ....
- ....
- ....
- ....
- ....
- ....
- ....
- ....

AFFIRMATION

....

....

INSPIRATION

....

....

My Morning Gratitude Notes

DATE

3 THINGS I AM GRATEFUL FOR

I AM LOOKING FORWARD TO

I AM AT MY HAPPIEST WHEN

HOW ARE YOU FEELING TODAY?

WHAT WOULD MAKE TODAY GREAT?

POSSITIVE QUOTE

ACTS OF KINDNESS I WILL DO TODAY

SELF CARE ACTIVITIES

My Evening Gratitude Notes

DATE

BEST MOMENTS OF THE DAY:

PEOPLE I'M GRATEFUL FOR:

WHAT WOULD HAVE MADE TODAY BETTER?

POSSITIVE AFFIRMATION OF THE DAY

THINGS I AM GRATEFUL FOR:

3 ACCOMPLISHMENTS FROM TODAY

THOUGHTS AND FEELINGS:

Date: ______________

TODAY'S MOOD

☹ ☹ 😐 ☺ 😃

THINGS THAT MADE ME HAPPY TODAY

1. ....
2. ....
3. ....
4. ....
5. ....

SELF-CARE LIST

- ....
- ....
- ....
- ....
- ....
- ....
- ....
- ....

AFFIRMATION

....
....

INSPIRATION

....
....

My Morning Gratitude Notes

DATE

3 THINGS I AM GRATEFUL FOR

WHAT WOULD MAKE TODAY GREAT?

I AM LOOKING FORWARD TO

POSSITIVE QUOTE

I AM AT MY HAPPIEST WHEN

ACTS OF KINDNESS I WILL DO TODAY

HOW ARE YOU FEELING TODAY?

SELF CARE ACTIVITIES

My Evening Gratitude Notes

DATE

BEST MOMENTS OF THE DAY:

PEOPLE I'M GRATEFUL FOR:

WHAT WOULD HAVE MADE TODAY BETTER?

POSSITIVE AFFIRMATION OF THE DAY

THINGS I AM GRATEFUL FOR:

3 ACCOMPLISHMENTS FROM TODAY

THOUGHTS AND FEELINGS:

Date: ______________

TODAY'S MOOD

😢 ☹ 😐 ☺ 😃

SELF-CARE LIST

- ....
- ....
- ....
- ....
- ....
- ....
- ....
- ....

THINGS THAT MADE ME HAPPY TODAY

1. ....
2. ....
3. ....
4. ....
5. ....

AFFIRMATION

....

....

INSPIRATION

....

....

My Morning Gratitude Notes

DATE

3 THINGS I AM GRATEFUL FOR

WHAT WOULD MAKE TODAY GREAT?

I AM LOOKING FORWARD TO

POSSITIVE QUOTE

I AM AT MY HAPPIEST WHEN

ACTS OF KINDNESS I WILL DO TODAY

HOW ARE YOU FEELING TODAY?

SELF CARE ACTIVITIES

My Evening Gratitude Notes

DATE

BEST MONTS OF THE DAY:

PEOPLE I'M GRATEFUL FOR:

WHAT WOULD HAVE MADE TODAY BETTER?

POSSITIVE AFFIRMATION OF THE DAY

THINGS I AM GRATEFUL FOR:

3 ACCOMPLISHMENTS FROM TODAY

THOUGHTS AND FEELINGS:

Date: ____________

TODAY'S MOOD

☹ ☹ 😐 ☺ 😃

THINGS THAT MADE ME HAPPY TODAY

1. ....
2. ....
3. ....
4. ....
5. ....

SELF-CARE LIST

- ....
- ....
- ....
- ....
- ....
- ....
- ....
- ....

AFFIRMATION

....

....

INSPIRATION

....

....

My Morning Gratitude Notes

DATE

3 THINGS I AM GRATEFUL FOR

WHAT WOULD MAKE TODAY GREAT?

I AM LOOKING FORWARD TO

POSSITIVE QUOTE

I AM AT MY HAPPIEST WHEN

ACTS OF KINDNESS I WILL DO TODAY

HOW ARE YOU FEELING TODAY?

SELF CARE ACTIVITIES

My Evening Gratitude Notes

DATE

BEST MOMENTS OF THE DAY:

PEOPLE I'M GRATEFUL FOR:

WHAT WOULD HAVE MADE TODAY BETTER?

POSSITIVE AFFIRMATION OF THE DAY

THINGS I AM GRATEFUL FOR:

3 ACCOMPLISHMENTS FROM TODAY

THOUGHTS AND FEELINGS:

Date: ____________

TODAY'S MOOD

☹ ☹ 😐 ☺ 😃

THINGS THAT MADE ME HAPPY TODAY

1. ....
2. ....
3. ....
4. ....
5. ....

SELF-CARE LIST

- ....
- ....
- ....
- ....
- ....
- ....
- ....
- ....

AFFIRMATION

....

....

INSPIRATION

....

....

My Morning Gratitude Notes

DATE

3 THINGS I AM GRATEFUL FOR

I AM LOOKING FORWARD TO

I AM AT MY HAPPIEST WHEN

HOW ARE YOU FEELING TODAY?

WHAT WOULD MAKE TODAY GREAT?

POSSITIVE QUOTE

ACTS OF KINDNESS I WILL DO TODAY

SELF CARE ACTIVITIES

My Evening Gratitude Notes

DATE

BEST MOMENTS OF THE DAY:

PEOPLE I'M GRATEFUL FOR:

WHAT WOULD HAVE MADE TODAY BETTER?

POSSITIVE AFFIRMATION OF THE DAY

THINGS I AM GRATEFUL FOR:

3 ACCOMPLISHMENTS FROM TODAY

THOUGHTS AND FEELINGS:

Date: ______________

TODAY'S MOOD

SELF-CARE LIST

THINGS THAT MADE ME HAPPY TODAY

1.
2.
3.
4.
5.

AFFIRMATION

INSPIRATION

My Morning Gratitude Notes

DATE

3 THINGS I AM GRATEFUL FOR

I AM LOOKING FORWARD TO

I AM AT MY HAPPIEST WHEN

HOW ARE YOU FEELING TODAY?

WHAT WOULD MAKE TODAY GREAT?

POSSITIVE QUOTE

ACTS OF KINDNESS I WILL DO TODAY

SELF CARE ACTIVITIES

My Evening Gratitude Notes

DATE

BEST MOMENTS OF THE DAY:

PEOPLE I'M GRATEFUL FOR:

WHAT WOULD HAVE MADE TODAY BETTER?

POSSITIVE AFFIRMATION OF THE DAY

THINGS I AM GRATEFUL FOR:

3 ACCOMPLISHMENTS FROM TODAY

THOUGHTS AND FEELINGS:

Date: ______________

TODAY'S MOOD

SELF-CARE LIST

-
-
-
-
-
-
-
-

THINGS THAT MADE ME HAPPY TODAY

1.
2.
3.
4.
5.

AFFIRMATION

INSPIRATION

My Morning Gratitude Notes

DATE

3 THINGS I AM GRATEFUL FOR

I AM LOOKING FORWARD TO

I AM AT MY HAPPIEST WHEN

HOW ARE YOU FEELING TODAY?

WHAT WOULD MAKE TODAY GREAT?

POSSITIVE QUOTE

ACTS OF KINDNESS I WILL DO TODAY

SELF CARE ACTIVITIES

My Evening Gratitude Notes

DATE

BEST MOMENTS OF THE DAY:

PEOPLE I'M GRATEFUL FOR:

WHAT WOULD HAVE MADE TODAY BETTER?

POSSITIVE AFFIRMATION OF THE DAY

THINGS I AM GRATEFUL FOR:

3 ACCOMPLISHMENTS FROM TODAY

THOUGHTS AND FEELINGS:

Date: ______________

TODAY'S MOOD

THINGS THAT MADE ME HAPPY TODAY

1. ....
2. ....
3. ....
4. ....
5. ....

SELF-CARE LIST

- ....
- ....
- ....
- ....
- ....
- ....
- ....
- ....

AFFIRMATION

....

....

INSPIRATION

....

....

My Morning Gratitude Notes

DATE

3 THINGS I AM GRATEFUL FOR

-
-
-

I AM LOOKING FORWARD TO

I AM AT MY HAPPIEST WHEN

HOW ARE YOU FEELING TODAY?

WHAT WOULD MAKE TODAY GREAT?

POSSITIVE QUOTE

ACTS OF KINDNESS I WILL DO TODAY

SELF CARE ACTIVITIES

My Evening Gratitude Notes

DATE

BEST MOMENTS OF THE DAY:

PEOPLE I'M GRATEFUL FOR:

WHAT WOULD HAVE MADE TODAY BETTER?

POSSITIVE AFFIRMATION OF THE DAY

THINGS I AM GRATEFUL FOR:

3 ACCOMPLISHMENTS FROM TODAY

THOUGHTS AND FEELINGS:

Date: ______________

TODAY'S MOOD

☹ ☹ 😐 ☺ 😃

SELF-CARE LIST

- ....
- ....
- ....
- ....
- ....
- ....
- ....
- ....

THINGS THAT MADE ME HAPPY TODAY

1. ....
2. ....
3. ....
4. ....
5. ....

AFFIRMATION

....

....

INSPIRATION

....

....

My Morning Gratitude Notes

DATE

3 THINGS I AM GRATEFUL FOR

I AM LOOKING FORWARD TO

I AM AT MY HAPPIEST WHEN

HOW ARE YOU FEELING TODAY?

WHAT WOULD MAKE TODAY GREAT?

POSSITIVE QUOTE

ACTS OF KINDNESS I WILL DO TODAY

SELF CARE ACTIVITIES

My Evening Gratitude Notes

DATE

BEST MOMENTS OF THE DAY:

PEOPLE I'M GRATEFUL FOR:

WHAT WOULD HAVE MADE TODAY BETTER?

POSSITIVE AFFIRMATION OF THE DAY

THINGS I AM GRATEFUL FOR:

3 ACCOMPLISHMENTS FROM TODAY

THOUGHTS AND FEELINGS:

Date: ______________

TODAY'S MOOD

☹ ☹ 😐 ☺ 😃

THINGS THAT MADE ME HAPPY TODAY

1. ....
2. ....
3. ....
4. ....
5. ....

SELF-CARE LIST

- ....
- ....
- ....
- ....
- ....
- ....
- ....
- ....

AFFIRMATION

....

....

INSPIRATION

....

....

My Morning Gratitude Notes

DATE

3 THINGS I AM GRATEFUL FOR

I AM LOOKING FORWARD TO

I AM AT MY HAPPIEST WHEN

HOW ARE YOU FEELING TODAY?

WHAT WOULD MAKE TODAY GREAT?

POSSITIVE QUOTE

ACTS OF KINDNESS I WILL DO TODAY

SELF CARE ACTIVITIES

My Evening Gratitude Notes

DATE

BEST MOMENTS OF THE DAY:

PEOPLE I'M GRATEFUL FOR:

WHAT WOULD HAVE MADE TODAY BETTER?

POSSITIVE AFFIRMATION OF THE DAY

THINGS I AM GRATEFUL FOR:

3 ACCOMPLISHMENTS FROM TODAY

THOUGHTS AND FEELINGS:

Date: ____________

TODAY'S MOOD

SELF-CARE LIST

- ....
- ....
- ....
- ....
- ....
- ....
- ....
- ....

THINGS THAT MADE ME HAPPY TODAY

1. ....
2. ....
3. ....
4. ....
5. ....

AFFIRMATION

....

....

INSPIRATION

....

....

My Morning Gratitude Notes

DATE

3 THINGS I AM GRATEFUL FOR

I AM LOOKING FORWARD TO

I AM AT MY HAPPIEST WHEN

HOW ARE YOU FEELING TODAY?

WHAT WOULD MAKE TODAY GREAT?

POSSITIVE QUOTE

ACTS OF KINDNESS I WILL DO TODAY

SELF CARE ACTIVITIES

My Evening Gratitude Notes

DATE

BEST MOMENTS OF THE DAY:

PEOPLE I'M GRATEFUL FOR:

WHAT WOULD HAVE MADE TODAY BETTER?

POSSITIVE AFFIRMATION OF THE DAY

THINGS I AM GRATEFUL FOR:

3 ACCOMPLISHMENTS FROM TODAY

THOUGHTS AND FEELINGS:

Date: ______________

TODAY'S MOOD

☹ ☹ 😐 ☺ 😀

THINGS THAT MADE ME HAPPY TODAY

1. ....

2. ....

3. ....

4. ....

5. ....

SELF-CARE LIST

- ....
- ....
- ....
- ....
- ....
- ....
- ....
- ....

AFFIRMATION

....

....

INSPIRATION

....

....

My Morning Gratitude Notes

DATE

3 THINGS I AM GRATEFUL FOR

I AM LOOKING FORWARD TO

I AM AT MY HAPPIEST WHEN

HOW ARE YOU FEELING TODAY?

WHAT WOULD MAKE TODAY GREAT?

POSSITIVE QUOTE

ACTS OF KINDNESS I WILL DO TODAY

SELF CARE ACTIVITIES

My Evening Gratitude Notes

DATE

BEST MOMENTS OF THE DAY:

PEOPLE I'M GRATEFUL FOR:

WHAT WOULD HAVE MADE TODAY BETTER?

POSSITIVE AFFIRMATION OF THE DAY

THINGS I AM GRATEFUL FOR:

3 ACCOMPLISHMENTS FROM TODAY

THOUGHTS AND FEELINGS:

Date: ____________

TODAY'S MOOD

SELF-CARE LIST

-
-
-
-
-
-
-
-

THINGS THAT MADE ME HAPPY TODAY

1.
2.
3.
4.
5.

AFFIRMATION

INSPIRATION

My Morning Gratitude Notes

DATE

3 THINGS I AM GRATEFUL FOR

I AM LOOKING FORWARD TO

I AM AT MY HAPPIEST WHEN

HOW ARE YOU FEELING TODAY?

WHAT WOULD MAKE TODAY GREAT?

POSSITIVE QUOTE

ACTS OF KINDNESS I WILL DO TODAY

SELF CARE ACTIVITIES

My Evening Gratitude Notes

DATE

BEST MOMENTS OF THE DAY:

PEOPLE I'M GRATEFUL FOR:

WHAT WOULD HAVE MADE TODAY BETTER?

POSSITIVE AFFIRMATION OF THE DAY

THINGS I AM GRATEFUL FOR:

3 ACCOMPLISHMENTS FROM TODAY

THOUGHTS AND FEELINGS:

Date: ____________

TODAY'S MOOD

☹ ☹ 😐 ☺ 😃

THINGS THAT MADE ME HAPPY TODAY

1. ....
2. ....
3. ....
4. ....
5. ....

SELF-CARE LIST

- ....
- ....
- ....
- ....
- ....
- ....
- ....
- ....

AFFIRMATION

....
....

INSPIRATION

....
....

My Morning Gratitude Notes

DATE

3 THINGS I AM GRATEFUL FOR

I AM LOOKING FORWARD TO

I AM AT MY HAPPIEST WHEN

HOW ARE YOU FEELING TODAY?

WHAT WOULD MAKE TODAY GREAT?

POSSITIVE QUOTE

ACTS OF KINDNESS I WILL DO TODAY

SELF CARE ACTIVITIES

My Evening Gratitude Notes

DATE

BEST MOMENTS OF THE DAY:

PEOPLE I'M GRATEFUL FOR:

WHAT WOULD HAVE MADE TODAY BETTER?

POSSITIVE AFFIRMATION OF THE DAY

THINGS I AM GRATEFUL FOR:

3 ACCOMPLISHMENTS FROM TODAY

THOUGHTS AND FEELINGS:

Date: ____________

TODAY'S MOOD

😢 ☹ 😐 ☺ 😃

THINGS THAT MADE ME HAPPY TODAY

1. ....
2. ....
3. ....
4. ....
5. ....

SELF-CARE LIST

- ....
- ....
- ....
- ....
- ....
- ....
- ....
- ....

AFFIRMATION

....

....

INSPIRATION

....

....

My Morning Gratitude Notes

DATE

3 THINGS I AM GRATEFUL FOR

I AM LOOKING FORWARD TO

I AM AT MY HAPPIEST WHEN

HOW ARE YOU FEELING TODAY?

WHAT WOULD MAKE TODAY GREAT?

POSSITIVE QUOTE

ACTS OF KINDNESS I WILL DO TODAY

SELF CARE ACTIVITIES

My Evening Gratitude Notes

DATE

BEST MONTS OF THE DAY:

PEOPLE I'M GRATEFUL FOR:

WHAT WOULD HAVE MADE TODAY BETTER?

POSSITIVE AFFIRMATION OF THE DAY

THINGS I AM GRATEFUL FOR:

3 ACCOMPLISHMENTS FROM TODAY

THOUGHTS AND FEELINGS:

Date: ____________

TODAY'S MOOD

☹ ☹ 😐 ☺ 😀

THINGS THAT MADE ME HAPPY TODAY

1. ....
2. ....
3. ....
4. ....
5. ....

SELF-CARE LIST

- ....
- ....
- ....
- ....
- ....
- ....
- ....
- ....

AFFIRMATION

....

....

INSPIRATION

....

....

My Morning Gratitude Notes

DATE

3 THINGS I AM GRATEFUL FOR

WHAT WOULD MAKE TODAY GREAT?

I AM LOOKING FORWARD TO

POSSITIVE QUOTE

I AM AT MY HAPPIEST WHEN

ACTS OF KINDNESS I WILL DO TODAY

HOW ARE YOU FEELING TODAY?

SELF CARE ACTIVITIES

My Evening Gratitude Notes

DATE

BEST MOMENTS OF THE DAY:

PEOPLE I'M GRATEFUL FOR:

WHAT WOULD HAVE MADE TODAY BETTER?

POSSITIVE AFFIRMATION OF THE DAY

THINGS I AM GRATEFUL FOR:

3 ACCOMPLISHMENTS FROM TODAY

THOUGHTS AND FEELINGS:

Date: ______________

TODAY'S MOOD

THINGS THAT MADE ME HAPPY TODAY

1. ....
2. ....
3. ....
4. ....
5. ....

SELF-CARE LIST

- ....
- ....
- ....
- ....
- ....
- ....
- ....
- ....

AFFIRMATION

....

....

INSPIRATION

....

....

My Morning Gratitude Notes

DATE

3 THINGS I AM GRATEFUL FOR

WHAT WOULD MAKE TODAY GREAT?

I AM LOOKING FORWARD TO

POSSITIVE QUOTE

I AM AT MY HAPPIEST WHEN

ACTS OF KINDNESS I WILL DO TODAY

HOW ARE YOU FEELING TODAY?

SELF CARE ACTIVITIES

My Evening Gratitude Notes

DATE

BEST MONTENTS OF THE DAY:

PEOPLE I'M GRATEFUL FOR:

WHAT WOULD HAVE MADE TODAY BETTER?

POSSITIVE AFFIRMATION OF THE DAY

THINGS I AM GRATEFUL FOR:

3 ACCOMPLISHMENTS FROM TODAY

THOUGHTS AND FEELINGS:

Date: ____________

TODAY'S MOOD

☹ ☹ 😐 ☺ 😃

THINGS THAT MADE ME HAPPY TODAY

1. ....
2. ....
3. ....
4. ....
5. ....

SELF-CARE LIST

- ....
- ....
- ....
- ....
- ....
- ....
- ....
- ....

AFFIRMATION

....

....

INSPIRATION

....

....

Date: ____________

TODAY'S MOOD

☹ ☹ 😐 ☺ 😀

THINGS THAT MADE ME HAPPY TODAY

1. ....
2. ....
3. ....
4. ....
5. ....

SELF-CARE LIST

- ....
- ....
- ....
- ....
- ....
- ....
- ....
- ....

AFFIRMATION

....

....

INSPIRATION

....

....

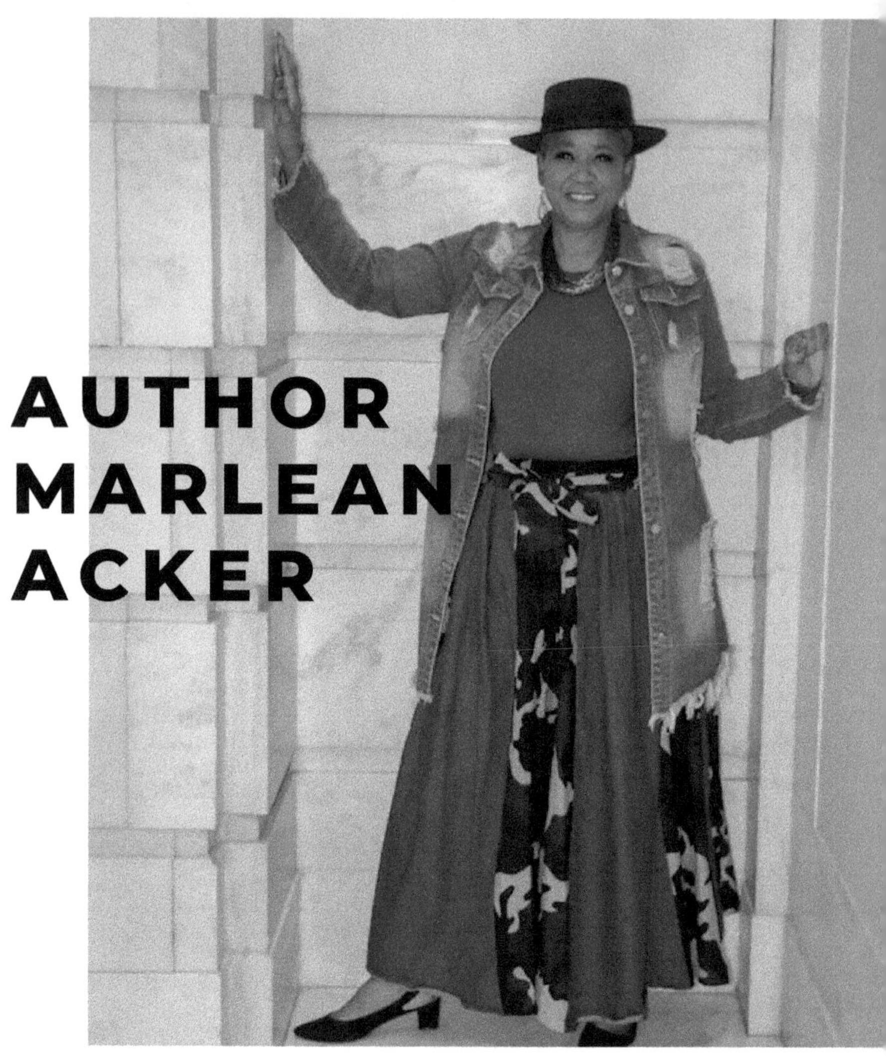

AUTHOR MARLEAN ACKER

When Every Day Is A New Day is a wonderful tribute to Author Marlean Acker's mom who was diagnosed with Dementia/Alzheimer. The journey she and her husband took with her mom as the disease progressed was hilarious, exhausting, and humbling. Acker's frankness and sense of gratitude (even amid tragedy) shines through and offers hope to anyone who has had to care for a loved one. While the primary message is for caregivers, this book encourages you to keep going, keep trying, and don't forget to give to yourself in the process.

Welcome

Founders of
Living Water
Books Publishing
Company

Visit our website
Livingwaterbooks.org

My husband Charles and I are the founders of Living Water Books Christian Publishing Company. I always knew from my childhood that I was chosen to record (write) for God. The prophetic gift rested upon my mother. My father told me I had the spiritual gift, but I needed to learn the skill, so he admonished me to pursue a higher education. I began college courses while in high school and three years later I received my B.A. Degree in Mass Communications with a concentration in News Editorial, Broadcasting, and Journalism. I met my husband and a few years later I released my first book, A Heart Unraveled, which I self-published. It became a best-seller allowing us to travel doing conferences, interviews, and book signings.

God established our business on the foundations of (John 7:38), Whoever believes in me, as scripture has said, rivers of living water will flow from within them. Living Water being symbolic for Holy Spirit living within us became the Living Waters within our writings as we prepared resources for our marriage ministry. Holy Spirit splashed through the pages and the testimonies that derived from the resources told us what we needed to do next. We looked at one another and said the name will be,

Living Water Books.

"Living Water Books is God's company. The Living Water of God into books distributed all over the world. God chose us as stewards and we are committed to serving God's Kingdom through God's people."

CONTACT US TODAY

WEBSITE: LIVINGWATERBOOKS.ORG

www.ingramcontent.com/pod-product-compliance
Lightning Source LLC
Chambersburg PA
CBHW060243030426
42335CB00014B/1578
* 9 7 9 8 9 8 6 8 2 8 6 4 0 *